W9-CEN-825

Genre Nonfiction

Essential Question
What are the properties of different types of elements?

Discovering the Elements

by Meish Goldish

Chapter 1
What Is an Element? 2

Chapter 2
Aluminum 5

Chapter 3
Iodine 8

Chapter 4
Helium 12

Chapter 5
The Periodic Table 15

Respond to Reading 20

PAIRED READ Breathless 21

Glossary 23

Index 24

Chapter 1

What Is an Element?

Everything around you is made of **matter**. Matter is anything that has weight and takes up space. Metal, wood, glass, and water are all matter.

People started to study matter long ago. They wondered why you can feel wind but not see it. They asked what water is and wanted to know why water could turn to ice. People had to learn more about matter so they could answer their questions.

People in ancient Greece thought that matter was made from the four materials of earth, water, fire, and air. They believed that everything had a different amount of these materials. A tree might have more fire than a rock.

ice cubes

Water was one of the ancient Greeks' four elements. Ice is water in solid form.

Modern scientists understand matter much better than people did in ancient Greece. We sort matter into the categories of solid, liquid, or a gas and know that matter is made of **elements**. Each element is pure, which means that it cannot be broken down into simpler parts.

Scientists now know that more than 100 different elements exist. People knew about some elements thousands of years ago and used elements such as iron and gold to make objects. However, most elements have been discovered only during the past 200 years.

categories groups

Solid

This book describes three of the elements and explains how they were discovered. Scientists learned a lot about matter when they discovered the elements. Many of the elements are important to us today.

You will also learn about a Russian chemist named Dmitri Mendeleyev (1834–1907). He organized the elements into a chart called the **periodic table** that is still used today. The chart groups together similar elements by their chemical properties.

Liquid and gas

teapot

Chapter 2

Aluminum

The element aluminum is the most common metal on Earth. There is a lot of it on Earth, but it was not easy to find. Scientists didn't find it until 1825. Why did it take so long?

Aluminum doesn't exist on its own. Together, aluminum and other elements form a **compound,** or <u>combination</u> of elements. The compound is different from the elements that made it. One compound is made of aluminum, sulfur, and oxygen. It helps dyes stick to paper.

Water is a common compound made from two elements that are gases—hydrogen and oxygen. Water can exist as a solid, liquid, or gas.

<u>combination</u> the joining together of

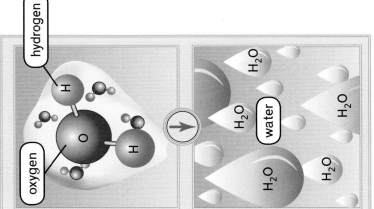

Water is a compound made from hydrogen and oxygen gases.

In the 1700s, scientists knew of a compound called alumina. They thought it was made from the element oxygen and another unknown element.

To find the other element, they had to break apart the compound. They didn't know how to do this.

In 1825, Danish chemist Hans Oersted (1777–1851) solved the problem. First, he created a new compound by mixing alumina with chlorine. Then he heated the new compound with other elements, creating a small amount of pure aluminum.

Oersted worked hard to find the new element. Later, scientists found easier ways to create large amounts of pure aluminum.

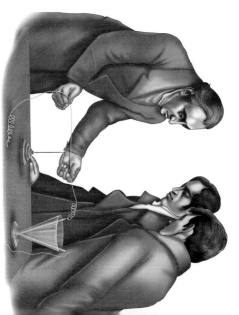

Oersted shows how he got aluminum from alumina.

Aluminum is cast into bars.

soda can

Used soda cans don't end up in the trash. They can be recycled.

Aluminum has many **properties** that make it useful. Pure aluminum is soft, making it easy to fold or roll.

An alloy is a mixture of aluminum and another metal. Alloys are very strong, so they are good for making buildings.

Aluminum will not rust in the rain. Builders use aluminum to make window frames and siding for houses.

Aluminum is also very light. Car manufacturers use aluminum to make light car parts. Light cars do not use a lot of gas. Many shippers use aluminum containers to move their goods because it costs less to ship light things.

Aluminum heats up quickly, which makes it useful for pots and pans. People can recycle aluminum, so many cans are made from aluminum. Aluminum is good for the environment.

manufacturers makers

Seaweed

Chapter 3

Iodine

Scientists did experiments to find aluminum and other elements in a lab, but some elements were discovered by chance. Iodine is an element discovered in this way in 1811.

The French scientist Bernard Courtois (1777–1838) was studying seaweed. He burned the seaweed and added sulfuric acid to the seaweed's ashes but accidentally poured on too much acid. A dark purple cloud of gas rose from the ashes. Then the gas settled on metal objects in the lab.

8

Dark iodine crystals

The gas cooled, and Courtois noticed that its form had changed. It was now a solid, not a gas. It had changed into shiny, dark **crystals**.

These crystals were a new element that Courtois had discovered by accident. He called it *iodine* because of its purple color. The Greek word *iodes* means "violet colored."

Today, scientists can make iodine in several ways. One way is to burn seaweed like Courtois did.

The discovery of iodine was important to medicine. The human body needs iodine for the thyroid gland to work properly. The thyroid allows the body to grow at a steady rate, and a lack of iodine slows down growth. Some people with thyroid problems get iodine from their doctors.

Humans need a small amount of iodine to survive, and missing this small amount of iodine can cause diseases like cancer. Young children need iodine to help their brains grow. Not enough iodine can also cause deafness. For these reasons, people who make salt often add iodine to it.

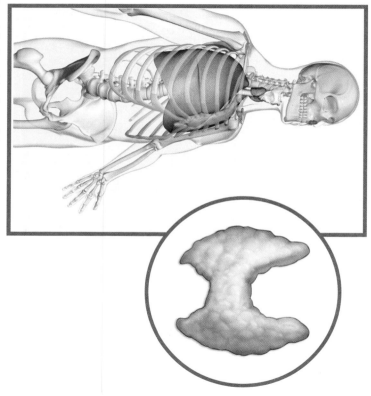

The thyroid gland is found in the neck. Your body gets iodine from different sources such as salt, milk, eggs, and seafood.

Iodine has other uses, too. Iodine kills bacteria. People add iodine to dirty water to make it safe to drink. If you have a cut, you can mix iodine and alcohol and put it on the cut to keep the cut from getting infected.

Iodine is also used for making and testing products. Iodine mixed with silver makes film for cameras. Chemists use iodine to test food for **starch**. Researchers place a drop of iodine on a food. The iodine turns dark blue if the food has a lot of starch. People also put iodine in some dyes and soaps.

An iodine stain helps make cells visible under a microscope.

iodine

Iodine kills germs on the skin.

Chapter 4

Helium

One day in 1868, Pierre Janssen, a French astronomer, was watching an eclipse of the Sun. He <u>observed</u> the Sun giving off a light wave that was a color he'd never seen before. Janssen knew that the Sun is made of different elements. Elements give off their own special colors when they are heated. The color helps you know what the element is. Janssen thought this strange new color must come from an element on the Sun.

The English scientist Sir Norman Lockyer agreed with Janssen. Lockyer named the new element *helium* after the Greek word *helios*, which means "Sun." Most scientists agreed that helium was on the Sun but did not think helium was on Earth. Then in 1895, Sir William Ramsay of Scotland discovered helium as he did tests on a metal called uranium.

observed saw for oneself

mine

Helium can be gotten from uranium mines like this one.

Helium is lighter than air. It is used to keep blimps flying.

Scientists experimented with helium and discovered its special properties. Helium is a gas that weighs very little. In fact, it is the second lightest of all the elements.

Helium is so light that gravity does not pull it toward Earth. The gas travels upward toward space. That is why helium is perfect inside blimps and birthday balloons. Helium lifts those items up in the air because it is lighter than the air surrounding it.

Helium also has medical uses. People with asthma take in a mixture of helium and oxygen through an inhaler to help them breathe. Helium is lighter than air so the mixture travels through the lungs more easily than air.

Deep-sea divers also use helium mixed with oxygen and other gases when they dive. They would feel a lot of pain deep under the water if they breathed normal air. The mixture of gases helps keep divers from getting sick.

At right is an inhaler for asthma. Below, helium helps divers breathe underwater.

oxygen tank

Chapter 5

The Periodic Table

Scientists had found 63 elements by 1869. They were solids, liquids, and gases, and all had different properties. They had different colors, and they reacted differently to heat and cold. They also <u>behaved</u> in different ways when mixed with other elements.

Scientists wondered about the elements. How were the elements alike? How were they different? Were they connected in any way?

Russian scientist Dmitri Mendeleyev found some answers. He began with 63 index cards. He wrote the names of the elements on the cards, with one element on a card. He wrote details about the elements on the cards.

<u>behaved</u> acted

A stamp from Russia shows Dmitri Medeleyev at work.

Mendeleyev included much information about **atoms**, or the smallest parts of an element. The atoms have different weights. For example, the atoms in aluminum weigh more than the atoms in helium. The weight of an atom in an element is called the **atomic weight**.

Mendeleyev wrote the atomic weight of each element on its card. Then he arranged the cards with the lightest atomic weights first and the ones with the heaviest atomic weights last.

Cu Copper

Ni Nickel

C Carbon

Co Cobalt

Mendeleyev laid out his 63 cards in a long row and looked for patterns. Soon he found one—every seventh card had something in common.

For example, his second card was a light metal called lithium. His ninth card was sodium, also a light metal. His sixteenth card, potassium, was a light metal, too. The cards in between these elements were not light metals.

Mendeleyev laid out the cards again. This time, he put them in rows of seven rather than in one long row. Then he read the columns from top to bottom. The elements in each column were alike in some ways. Mendeleyev was excited. The elements had some order!

Hg Mercury

Pb Lead

Fe Iron

Au Gold

Mendeleyev called his order of the elements the periodic table. He saw that some columns had places where elements didn't seem to fit, but he did not try to fill in the spaces. He thought they would be filled when people discovered new elements. That is exactly what happened!

Mendeleyev's periodic table grouped the elements.

Period	Group I	II	III	IV	V	VI	VII	VIII
1	H=1							
2	Li=7	Be=9.4	B=11	C=12	N=14	O=16	F=19	
3	Na=23	Mg=24	Al=27.3	Si=28	P=31	S=32	Cl=35.5	
4	K=39	Ca=40	?=44	Ti=48	V=51	Cr=52	Mn=55	Fe=56, Co=59 Ni=59
5	Cu=63	Zn=65	?=68	?=72	As=75	Se=78	Br=80	
6	Rb=85	Sr=87	?Yt=88	Zr=90	Nb=94	Mo=96	?=100	Ru=104, Rh=104 Pd=106
7	Ag=108	Cd=112	In=113	Sn=118	Sb=122	Te=125	J=127	
8	Cs=133	Ba=137	?Di=138	?Ce=140				
9								
10			?Er=178	?La=180	Ta=182	W=184		Os=195, Ir=197 Pt=198
11	Au=199	Hg=200	Tl=204	Pb=207	Bi=208			
12				Th=231		U=240		

Scientists went on to discover 90 elements that occur in nature. Then they kept looking. Scientists have made more in the laboratory. They don't make one after the next. Instead they look for patterns as Mendeleyev did. They place a new element in the right position on the table based on its properties.

Today's periodic table looks a bit different from the original periodic table. Even so, today's periodic table is based on Mendeleyev's work. His work was brilliant!

Periodic Table of the Elements

Group 1	2	3	4	5	6	7	8	9	10	11	12	13	14	15	16	17	18
1 H 1.008																	2 He 4.003
3 Li 6.941	4 Be 9.012											5 B 10.81	6 C 12.01	7 N 14.01	8 O 16	9 F 19	10 Ne 20.18
11 Na 22.99	12 Mg 24.31											13 Al 26.98	14 Si 28.09	15 P 30.97	16 S 32.07	17 Cl 35.45	18 Ar 39.95
19 K 39.10	20 Ca 40.08	21 Sc 44.96	22 Ti 47.88	23 V 50.94	24 Cr 52	25 Mn 54.94	26 Fe 55.85	27 Co 58.47	28 Ni 58.69	29 Cu 63.55	30 Zn 65.39	31 Ga 69.72	32 Ge 72.59	33 As 74.92	34 Se 78.96	35 Br 79.9	36 Kr 83.8
37 Rb 85.47	38 Sr 87.62	39 Y 88.91	40 Zr 91.22	41 Nb 92.91	42 Mo 95.94	43 Tc (98)	44 Ru 101.1	45 Rh 102.9	46 Pd 106.4	47 Ag 107.9	48 Cd 112.4	49 In 114.8	50 Sn 118.7	51 Sb 121.8	52 Te 127.6	53 I 126.9	54 Xe 131.3
55 Cs 132.9	56 Ba 137.3	57 La 138.9	72 Hf 178.5	73 Ta 180.9	74 W 183.9	75 Re 186.2	76 Os 190.2	77 Ir 192.2	78 Pt 195.1	79 Au 197	80 Hg 200.5	81 Tl 204.4	82 Pb 207.2	83 Bi 209	84 Po (210)	85 At (210)	86 Rn (222)
87 Fr (223)	88 Ra (226)	89 Ac (227)	104 Rf (257)	105 Db (260)	106 Sg (263)	107 Bh (262)	108 Hs (265)	109 Mt (266)	110 Ds (271)	111 Rg (272)	112 Uub (285)	113 Uut (284)	114 Uuq (289)	115 Uup (288)	116 Uuh (292)	117 Uus 0	118 Uuo 0

58 Ce 140.1	59 Pr 140.9	60 Nd 144.2	61 Pm (147)	62 Sm 150.4	63 Eu 152	64 Gd 157.3	65 Tb 158.9	66 Dy 162.5	67 Ho 164.9	68 Er 167.3	69 Tm 168.9	70 Yb 173	71 Lu 175
90 Th 232	91 Pa (231)	92 U (238)	93 Np (237)	94 Pu (242)	95 Am (243)	96 Cm (247)	97 Bk (247)	98 Cf (249)	99 Es (254)	100 Fm (253)	101 Md (256)	102 No (254)	103 Lr (257)

The modern Periodic Table of the Elements

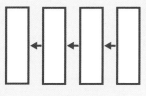

Respond to Reading

Summarize

Use important details from *Discovering the Elements* to summarize the selection. Your graphic organizer may help you.

Text Evidence

1. In what ways are aluminum and helium different from each other?

2. Make a sequence chart that shows the steps in the process that Mendeleyev used to make the periodic table. Be sure to list the steps in order. SEQUENCE

3. Reread pages 8 and 9. Explain why *iodine* is a good name for the element discovered by Bernard Courtois. GREEK ROOTS

4. Use various resources to find out how one element was discovered. Use a sequence chart to write an essay. In your essay, explain the process that was followed to discover that element. Share your writing with a partner. WRITE ABOUT READING

Compare Texts

Read how one of Earth's elements helped Namitha breathe better.

BREATHLESS

It was a hot, humid day in mid-July. Sara was hiking along a mountain trail with her friend Namitha.

Sara wiped a drop of sweat from her forehead. She glanced over at her friend Namitha, who had a hard time with the heat. Sara said, "It sure is steamy outside."

Namitha asked, "Do you mind if we sit down and rest for a moment?"

Namitha sat on a boulder and took off her heavy backpack. She went through it and removed a bottle of water and a small L-shaped container. Sara watched as Namitha placed the L-shaped container on her mouth. Namitha pressed a small button and inhaled. Then she sipped some water and rinsed her mouth with it.

Sara stared at her friend and wondered what she was doing.

Namitha held the L-shaped object in her hands for Sara to see. She explained that it was an inhaler that allowed her to breathe better on humid days.

Namitha said, "I have a condition called asthma that can tighten my airways and make breathing harder."

She explained that humid conditions make the condition worse and cause her to use the inhaler. Namitha explained that her asthma was once so bad that she went to the hospital and was treated with helium and oxygen.

"Is it just like the helium in a balloon?" Sara asked.

Namitha nodded and said that helium gas was one of the lightest gases in the air. She added that this made it easier for her to breathe. "Balloons and I have a lot in common because we may rely on helium to help us get around."

Make Connections

TEXT TO TEXT

How did helium help Namitha breathe better?

Glossary

atom (*AT-uhm*) the smallest unit of an element that keeps the properties of that element (*page 16*)

atomic weight (*uh-TOM-ik WAYT*) the weight of one atom of an element (*page 16*)

compound (*KOM-pownd*) any substance that is formed by the chemical combination of two or more elements held together by chemical bonds that cannot be separated by physical means; a compound has properties unlike those of the elements that make up the compound (*page 5*)

crystal (*KRIS-tuhl*) the geometric shape a mineral forms when its atoms and molecules get into fixed patterns (*page 9*)

element (*EL-uh-muhnt*) a pure substance that cannot be broken down into any simpler substances (*page 3*)

matter (*MAT-uhr*) a solid, liquid, or gas; anything that has mass or volume (*page 2*)

periodic table (*peer-ee-OD-ik TAY-buhl*) a chart that arranges all of the chemical elements (*page 4*)

property (*PROP-uhr-tee*) a characteristic of matter that can be observed, such as mass, volume, weight, or density (*page 7*)

starch (*STAHRCH*) a white food substance made and stored in most plants; it has no taste or smell; potatoes, wheat, corn, and rice have starch (*page 11*)

Index

alloy, 7

alumina, 6

aluminum, 5–8, 16

atom, 16

atomic weight, 16

compound, 5–6

Courtois, Bernard, 8–9

crystal, 9

eclipse, 12

element, 3–6, 8–9, 11–13, 15–18

gas, 3, 5, 7–9, 13, 15

helium, 12–14, 16

hydrogen, 5

iodine, 8–11, 14

Janssen, Pierre, 12

liquid, 3, 15

Lockyer, Sir Norman, 12

matter, 2–4

Mendeleyev, Dmitri, 4, 15–19

metal, 2, 5, 7, 8, 12, 17

Oersted, Hans, 6

oxygen, 5–6, 14

periodic table, 4, 15, 18–19

properties, 7

Ramsay, Sir William, 12

seaweed, 8–9

solid, 2–3, 9, 15

starch, 11

thyroid, 10

uranium, 12

water, 2, 5, 11, 14